Science Activity Books
ARTHROPODS

Created by The Good and the Beautiful Team

Cover Illustrated by Sandra Eide
Pages Illustrated by Zoe Damoulakis
Design by Phillip Colhouer

© 2024 The Good and the Beautiful, LLC
goodandbeautiful.com

Lesson 1

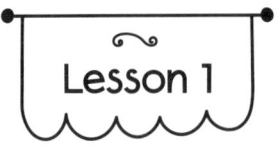

Draw lines to connect the dots on the spider and scorpion. Color the creatures.

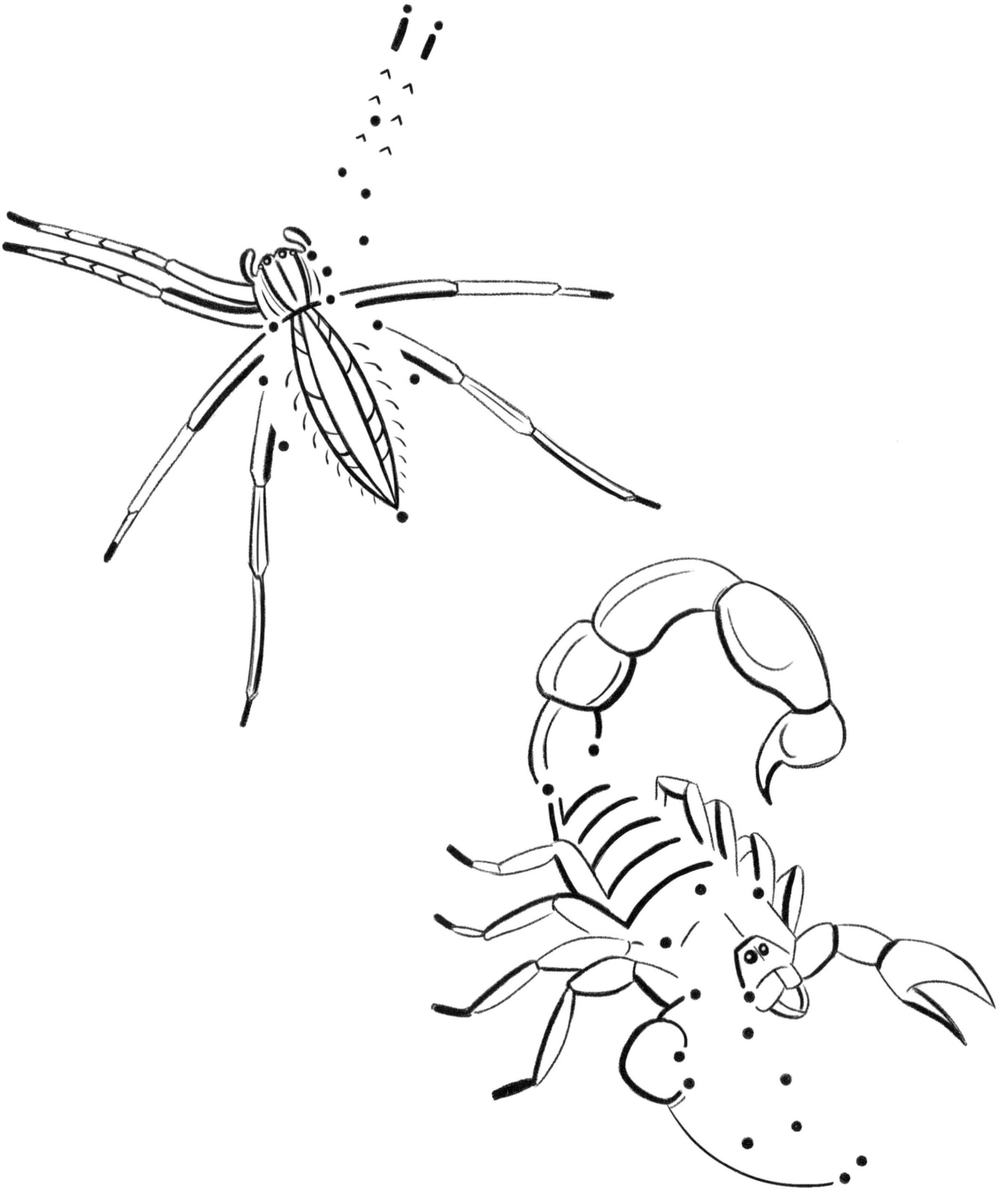

Cross out the items in the small boxes as you find them in each of the large scenes. Color the forest and underwater scenes.

Lesson 2

Draw lines to connect the matching ladybugs. Color the ladybugs.

Color sections with a 1 blue. Color sections with a 2 red. Color sections with a 3 green. Parent tip: Color the images of the crayons the correct colors to guide the child.

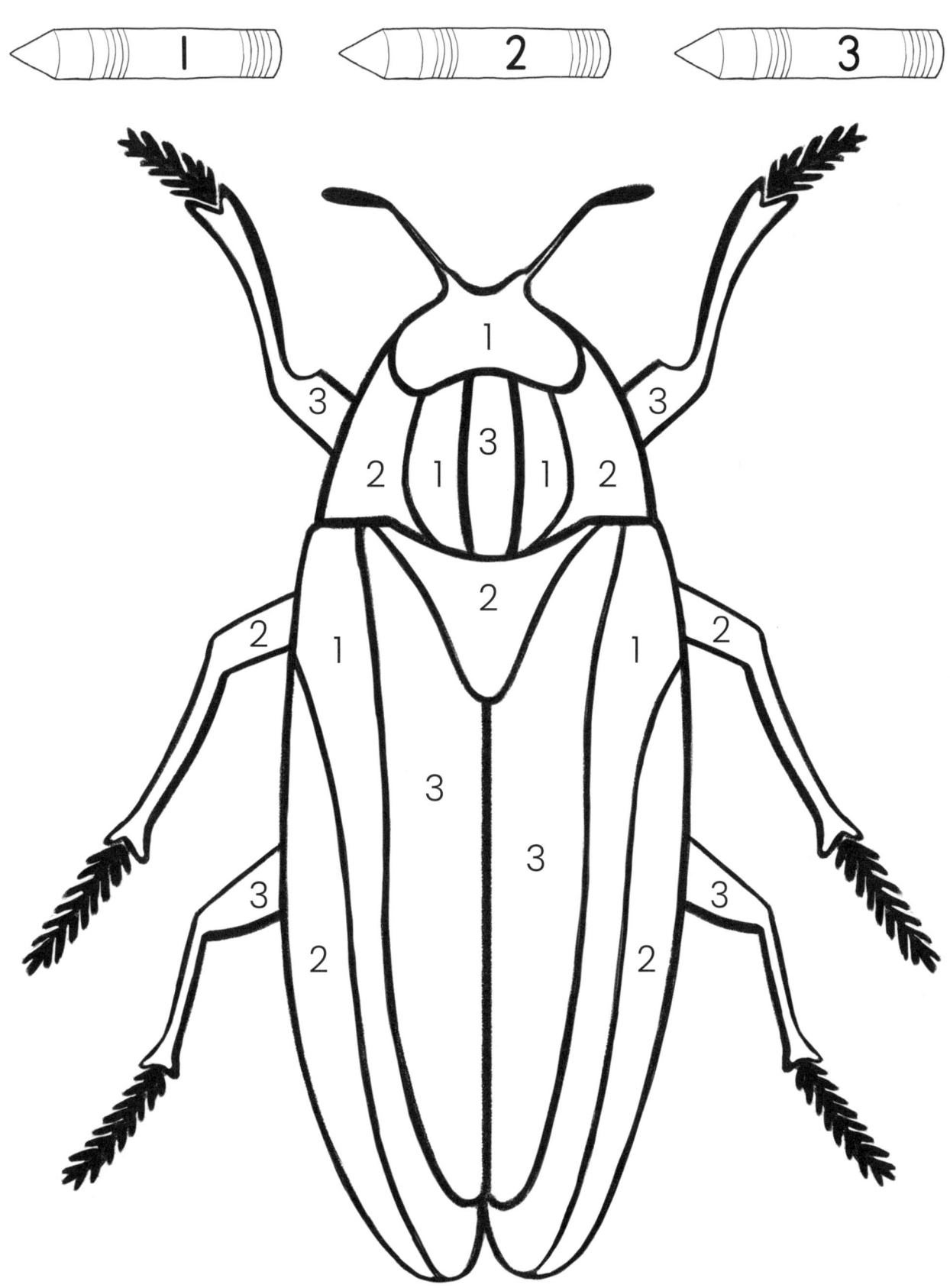

Lesson 3

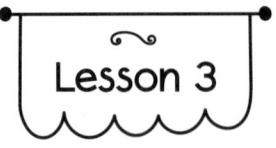

Circle the six items in the top picture that are not in the bottom picture. Color the top picture.

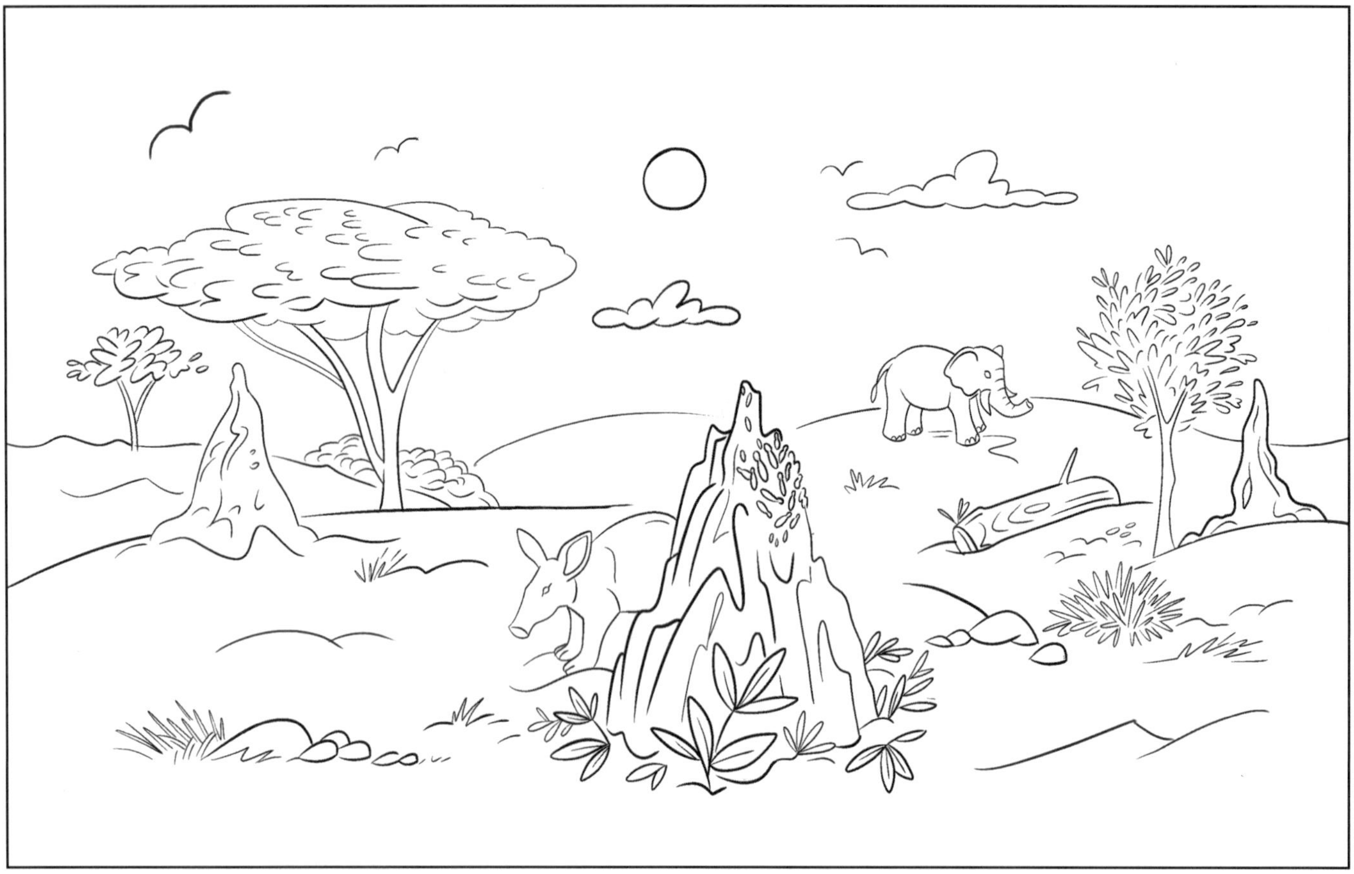

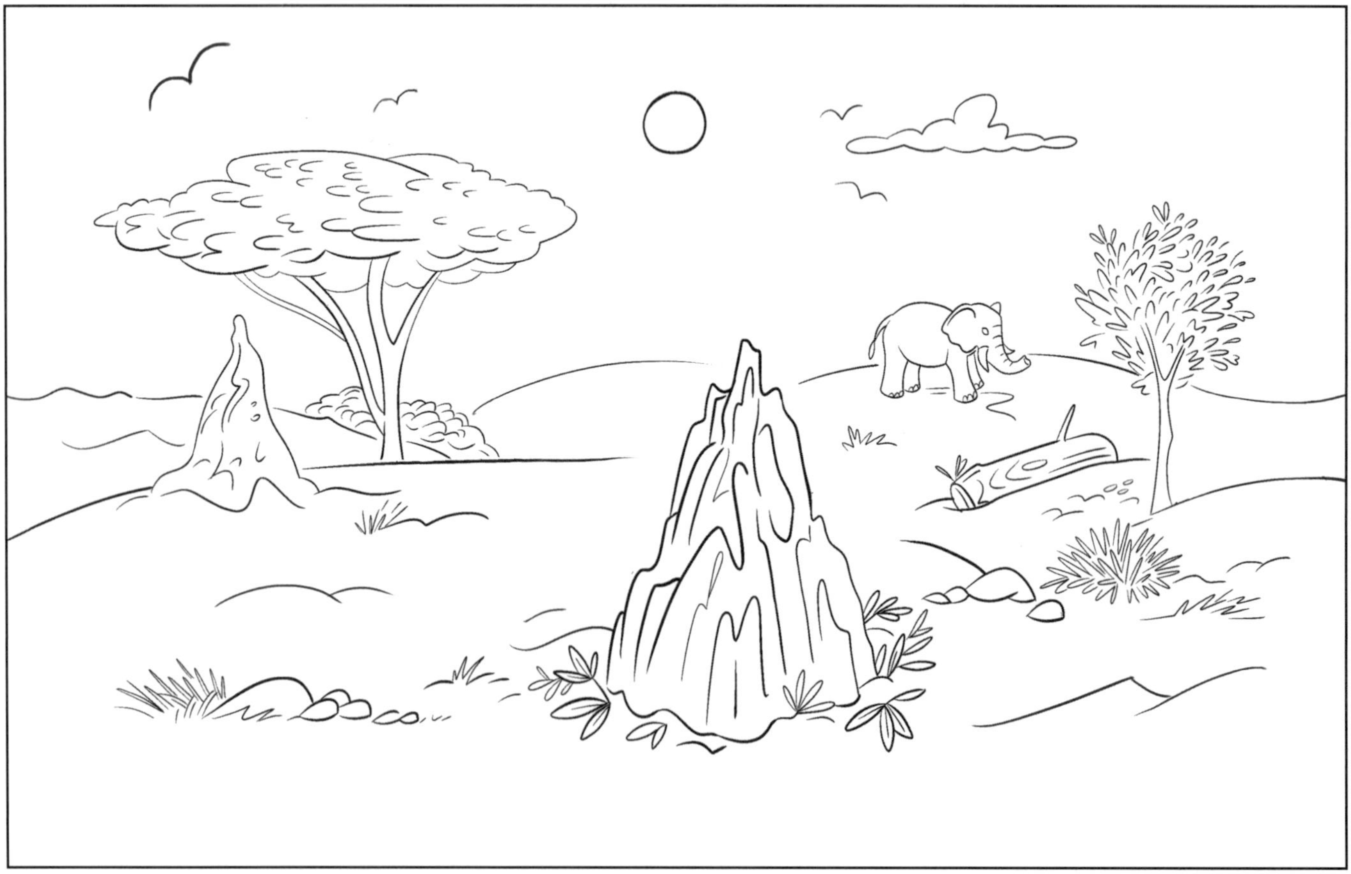

Complete the maze. Start at the arrow, draw over the silkworm cocoons, and finish at the silk moth. Color the cocoons and moth.

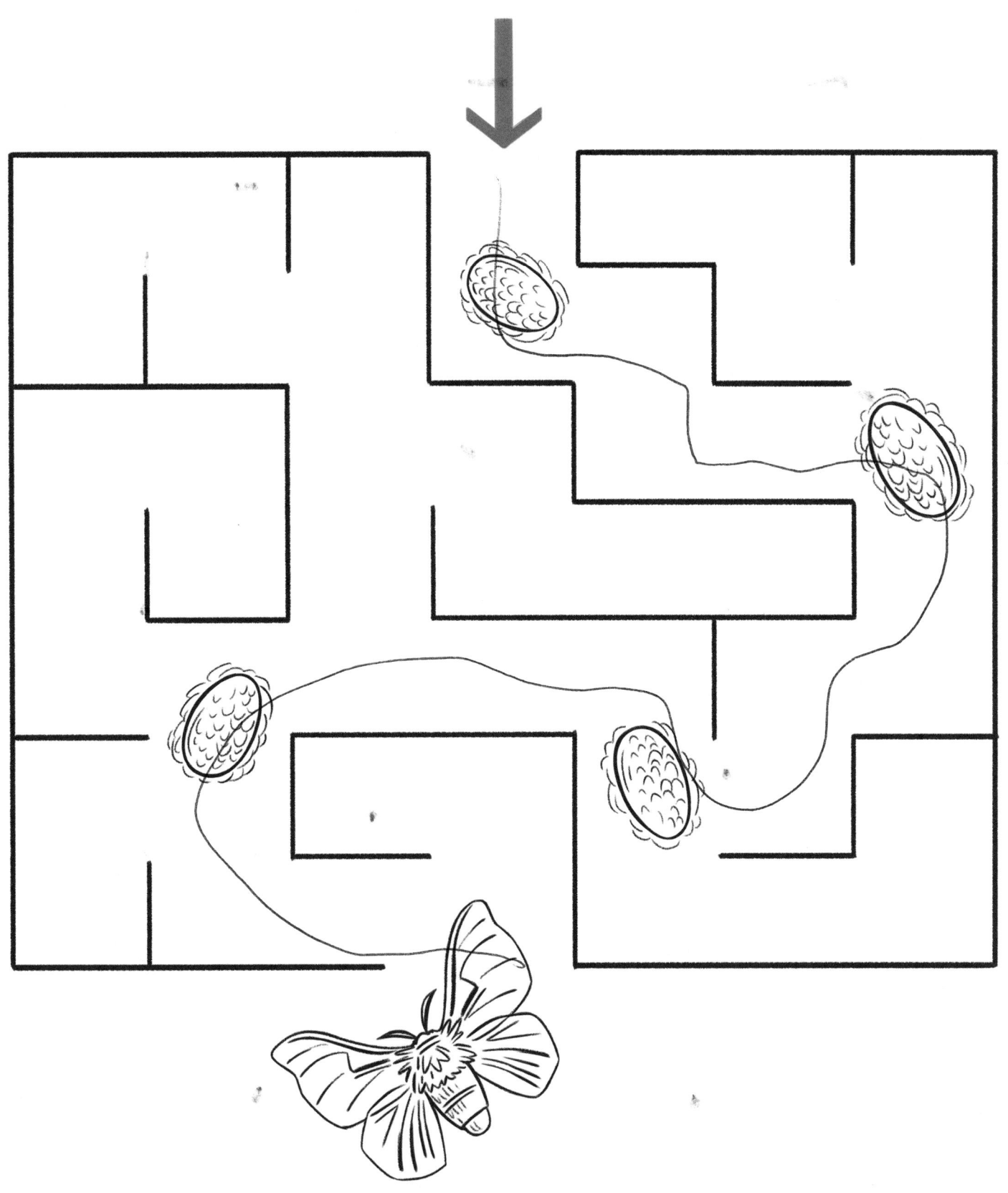

Lesson 4

Color the 10 food items below. We have these foods because bees pollinate their plants.

Draw lines to connect the separate honeycomb cells. Color the bee larva, worker bee, queen bee, and honey.

Lesson 5

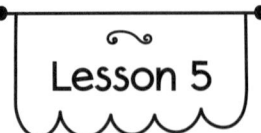

Trace the lines to connect each chrysalis on the left to the butterfly that will come out of it on the right. Color the pictures.

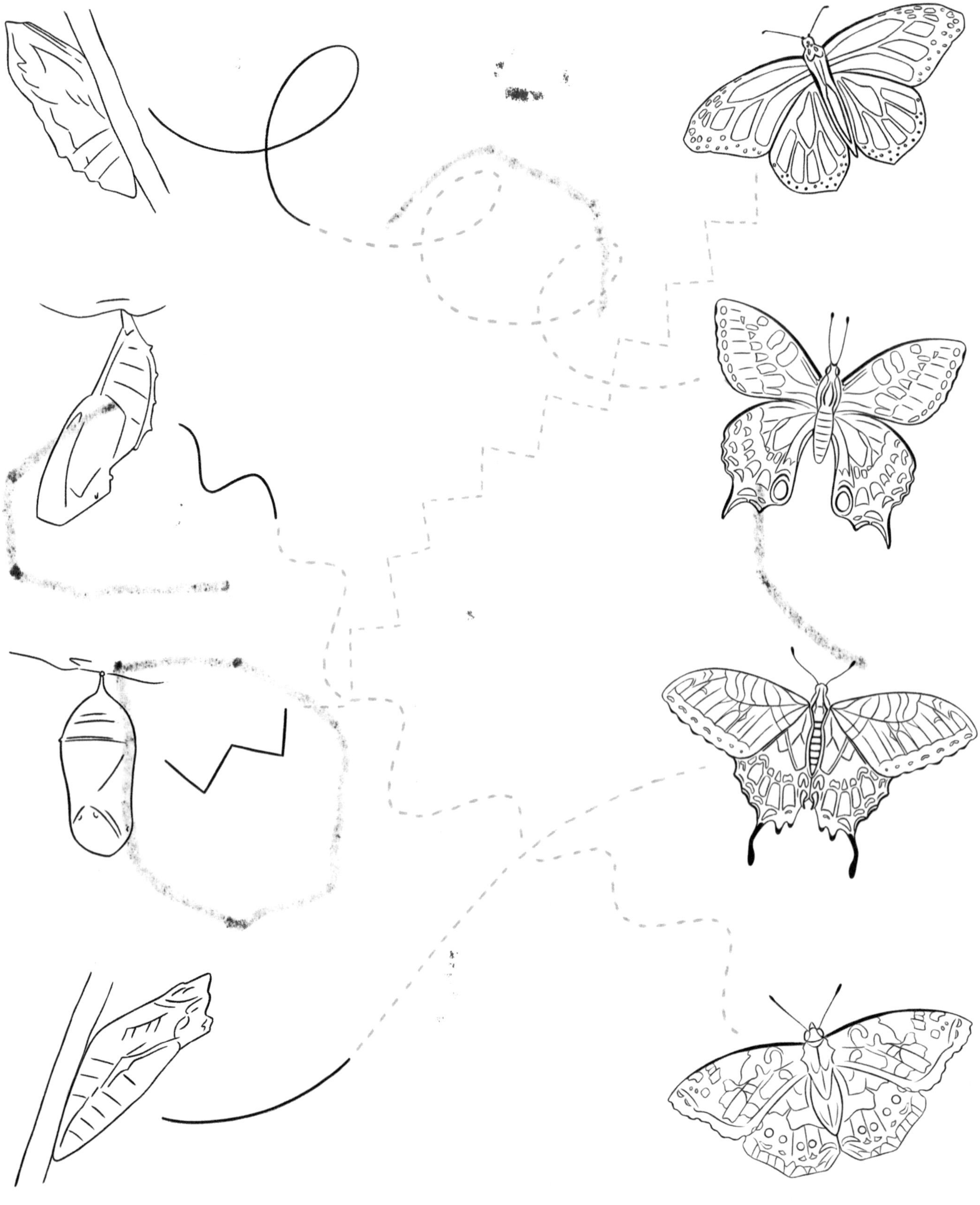

Color the number of creatures listed at the beginning of each row.

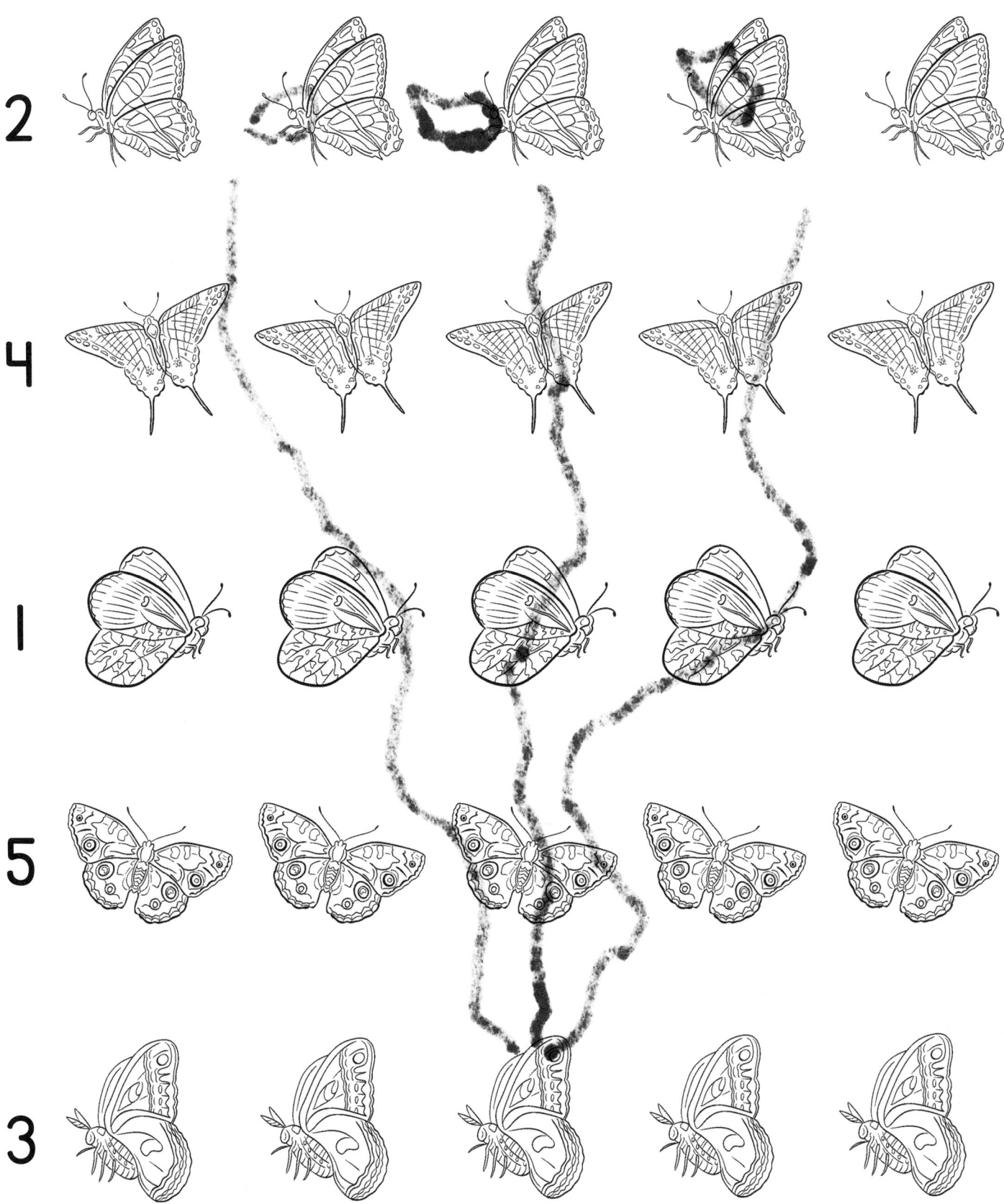

Lesson 6

Trace the lines from the leaf-cutter ants to their nest. Trace the leaves held by each ant. Color the picture.

Color or circle only the ants.

Lesson 7

The caterpillar below has many defenses. Some of them are shown in the circle below. Color the defenses you see on the caterpillar. Color the caterpillar.

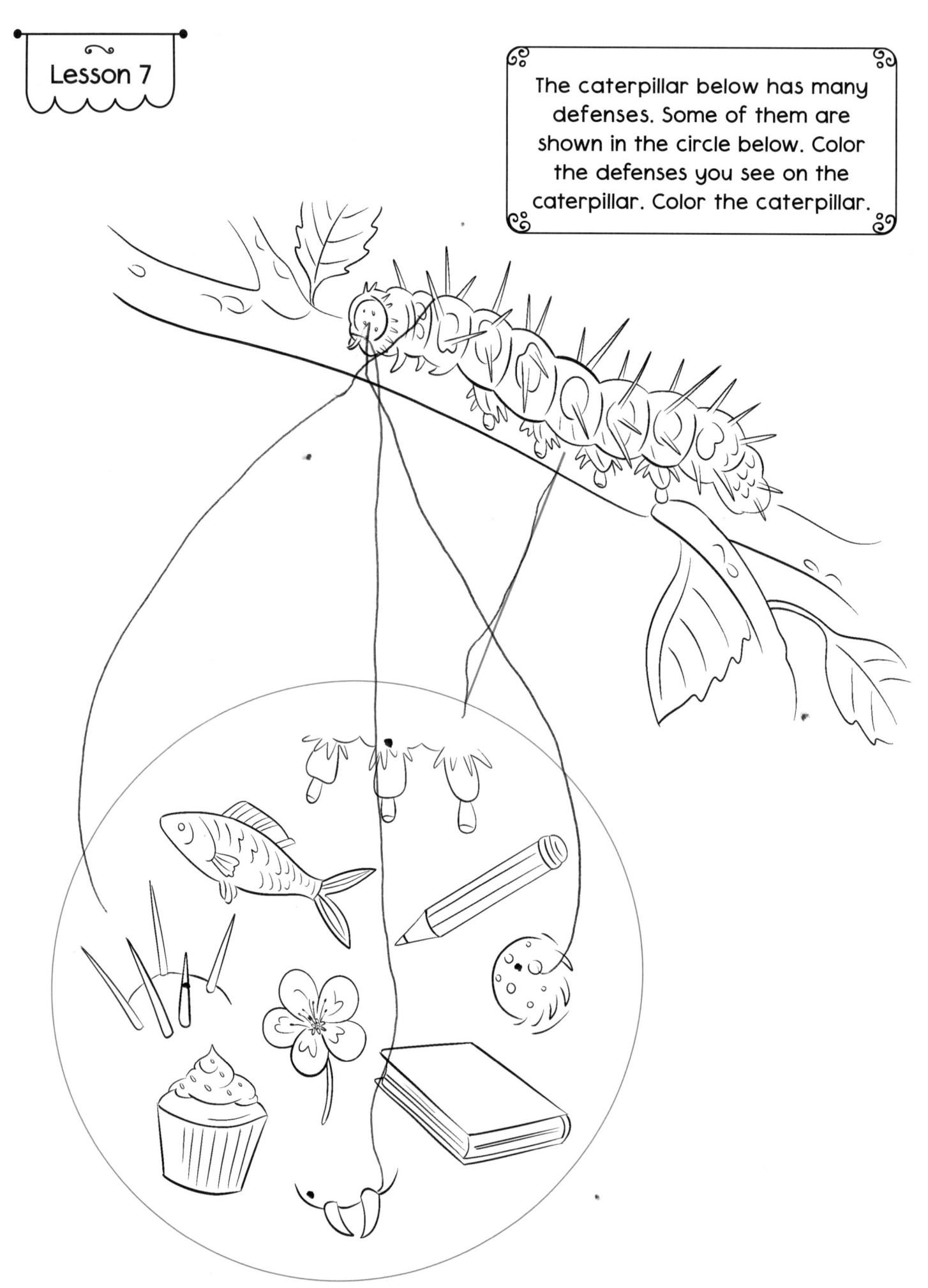

Draw a line from each creature to its matching shadow. Color the creatures.

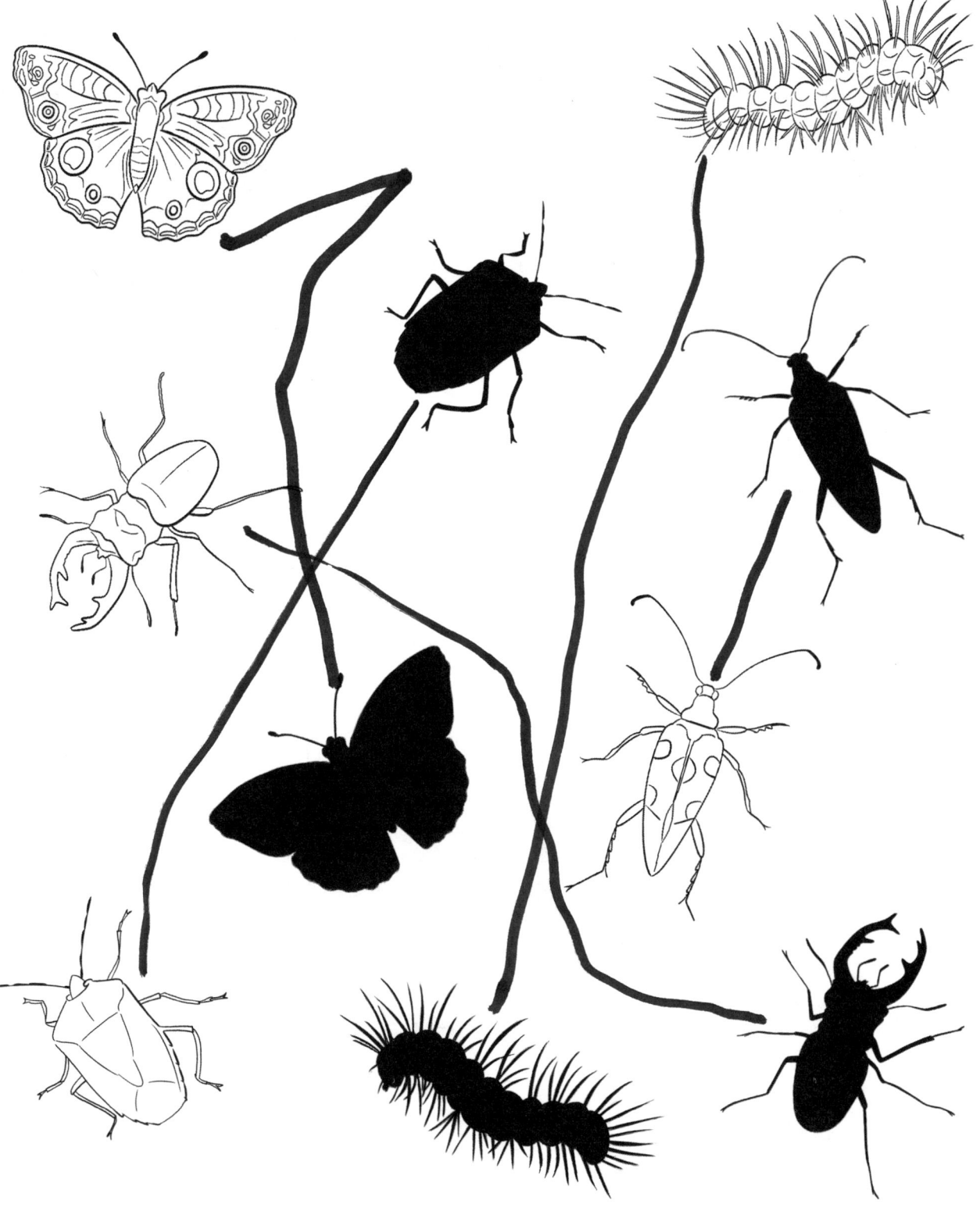

Trace the gray lines to complete the spider's web. Color the creatures. Cut out the creatures and tape or paste them onto the web.

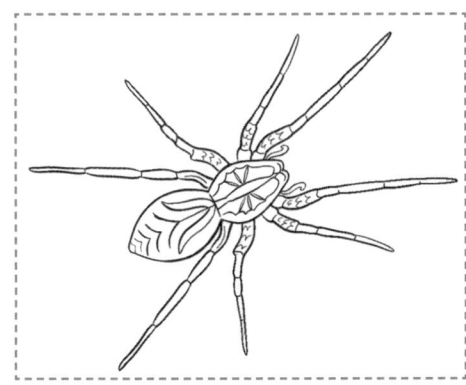

✂ Page intentionally left blank ✄

Page intentionally left blank

Lesson 9

Cut out the puzzle pieces. Arrange the puzzle correctly. Color the pieces and tape or paste them together on the top half of this page.

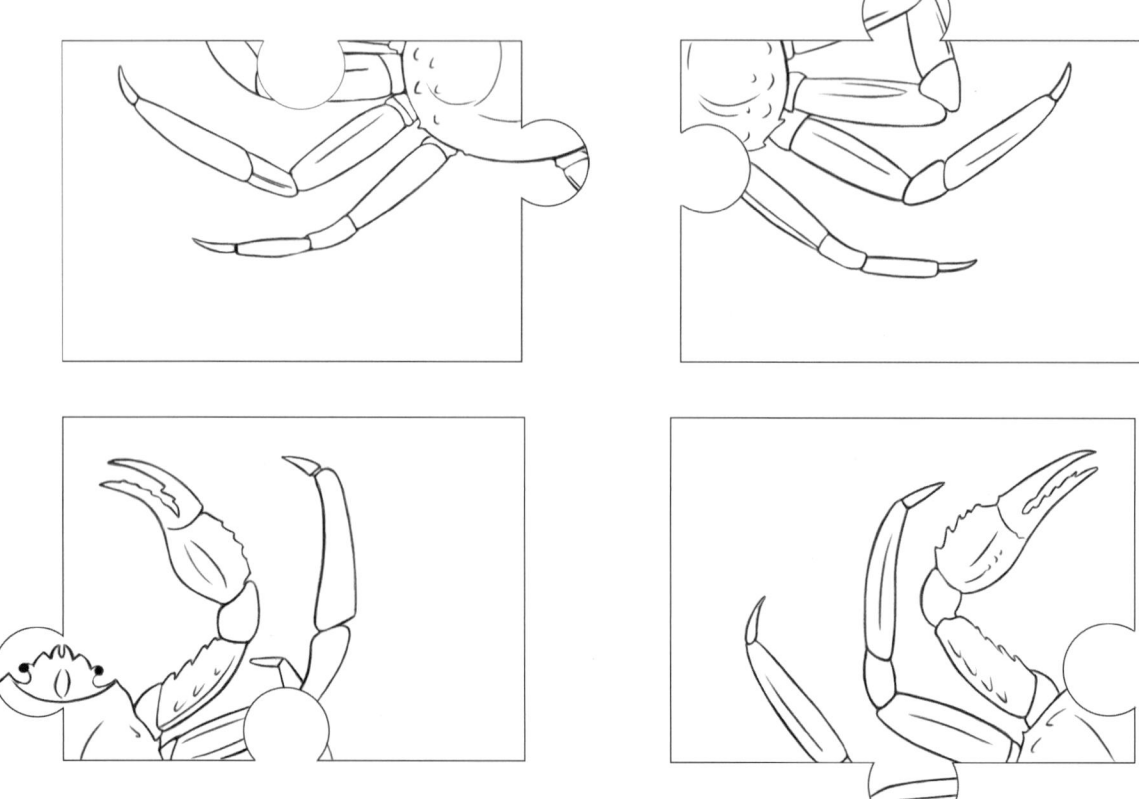

Trace the words "pill bug" and "shrimp." Color the creatures.

pill bug

shrimp

Lesson 10

Color the creature that exactly matches the first in each row.

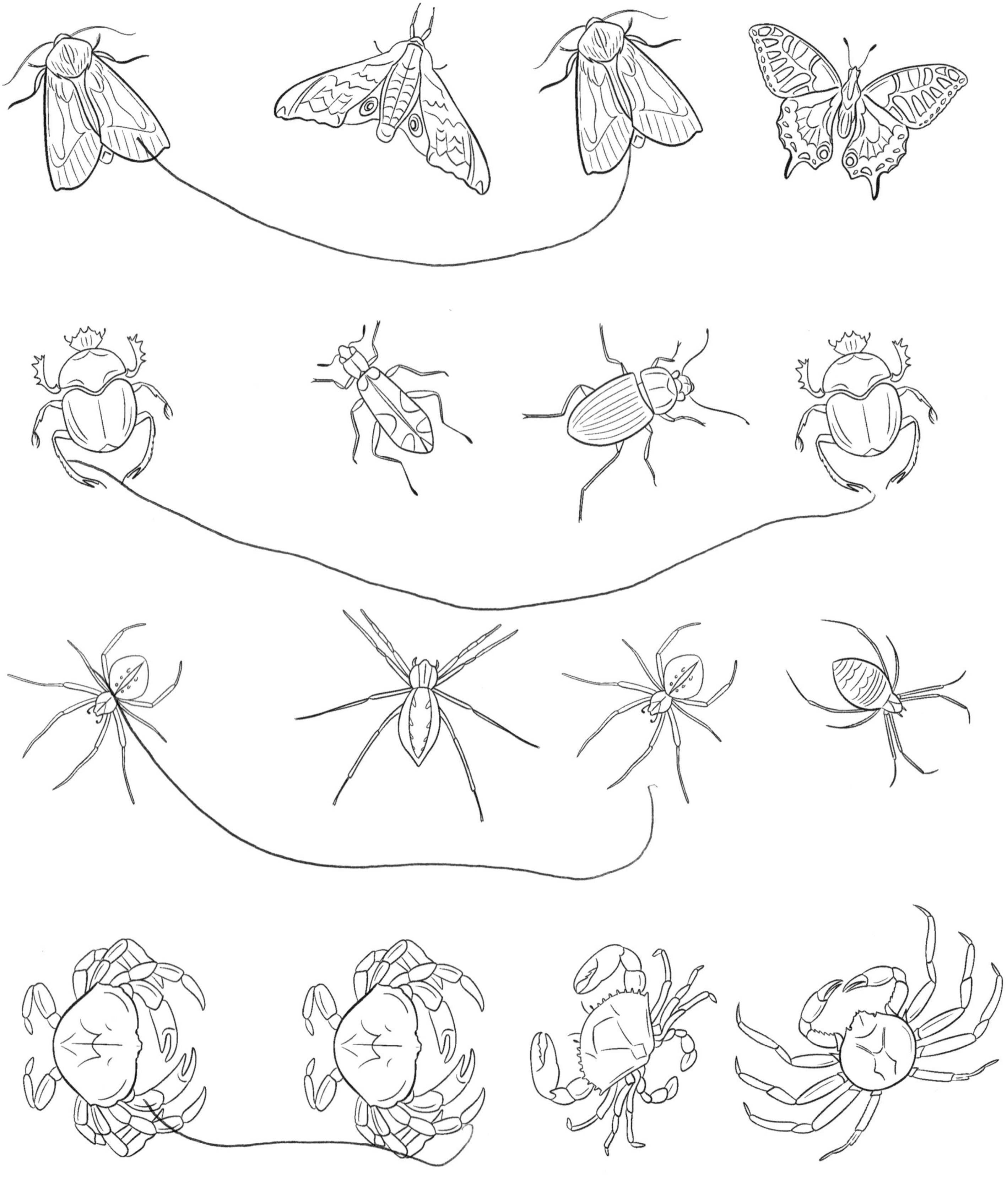

Extra Doodling and Drawing Page

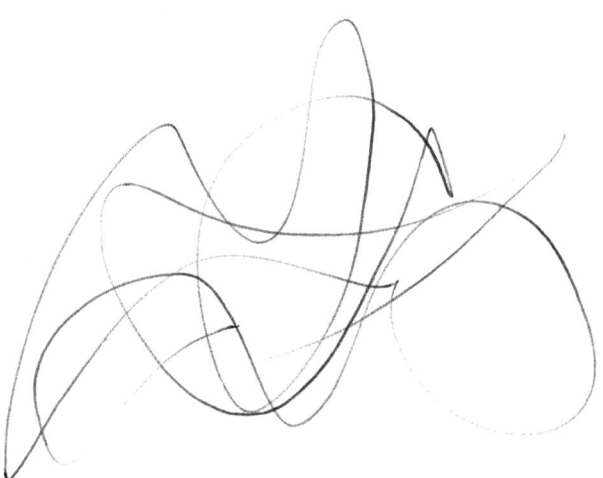